For Ben and Lawrie – C.D.

Guy, Guy, cherry pie – L.L.

A Red Fox Book

Published by Random House Children's Books
20 Vauxhall Bridge Road, London SW1V 2SA

A division of Random House UK Ltd
London Melbourne Sydney Auckland
Johannesburg and agencies throughout the world

1 3 5 7 9 10 8 6 4 2

First published by The Bodley Head Children's Books 1993

Red Fox edition 1996

Printed in China

RANDOM HOUSE UK Limited Reg. No. 954009

ISBN 0 09 923581 1

What is the Moon?

What
is the
Moon?

by Caroline Dunant
Pictures by Liz Loveless

Red Fox

What is the moon?

It's a light in the sky
that makes the dark bright.

Does it shine every night?

Sometimes we can't see it
but the moon is still there.

But where is it, where?

It might go behind clouds,
maybe to sleep, or to
play hide and seek.

Will it play with me?

We'll see, but now it's
time for your tea.

Why is the moon so high?
What's it doing up there?

At the top of the world, like the top of
a tree, the moon hangs by a ribbon
that no one can see.

What colour is it?
Why doesn't it show?
And how do you know?

I was told it is black, a long time ago.

Can I fly up and see?

Not just now. It's time for your tea.

Is the moon very old?
Is it hot, is it cold?

The moon changes shape, but it has
always been there. You would need
a coat, or something warm to wear.

What shapes do you mean?
Can you please show me?

I will, and then it's time for your tea.
Now, hold up your hands and we will
try to make shapes of the moon
changing in the sky.

Into the curve of the right,
fits the moon when it's new.

Make a circle with both and
the full moon's in view.

Then curve the left, to fit the
moon when it's old. And that
is what I was always told.

Can you do that again?
Can I try?
Can I see how the moon fits together with my hands and me?

Just once more, then you must have your tea.

But there's still so much
that I don't know.
Is the moon made of snow?

There are mountains and
valleys and wide open spaces.

It looks too small to have so many places.

The moon is a world as big as ours but no people are there, it just lives with the stars.

*That sounds very sad. I'll fly up and see
if the moon would like to play with me.*

No, not right now, it's so far away,
to get there would take a year and a day.

But look at the sky, the moon is so near.

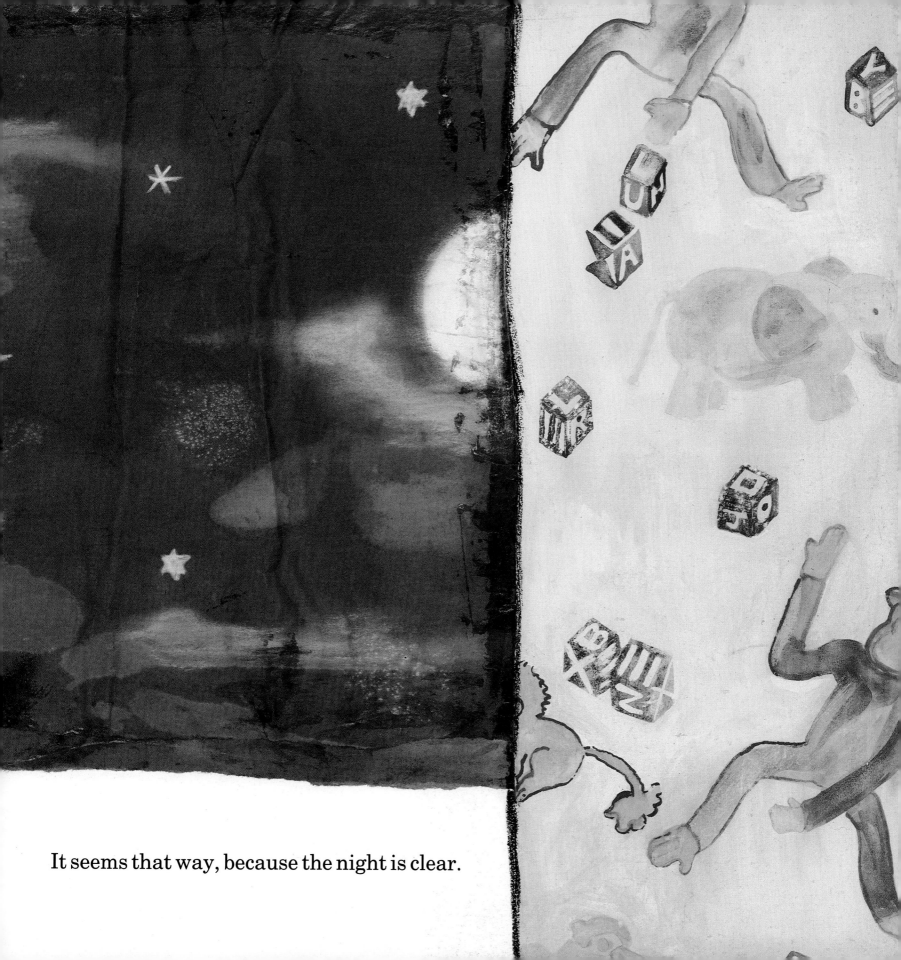

It seems that way, because the night is clear.

But Mum, can't you see, it's waving to me . . .
I'm going right now to the moon for my tea.

Some bestselling Red Fox picture books

THE BIG ALFIE AND ANNIE ROSE STORYBOOK
by Shirley Hughes
OLD BEAR
by Jane Hissey
OI! GET OFF OUR TRAIN
by John Burningham
DON'T DO THAT!
by Tony Ross
NOT NOW, BERNARD
by David McKee
ALL JOIN IN
by Quentin Blake
THE WHALES' SONG
by Gary Blythe and Dyan Sheldon
JESUS' CHRISTMAS PARTY
by Nicholas Allan
THE PATCHWORK CAT
by Nicola Bayley and William Mayne
MATILDA
by Hilaire Belloc and Posy Simmonds
WILLY AND HUGH
by Anthony Browne
THE WINTER HEDGEHOG
by Ann and Reg Cartwright
A DARK, DARK TALE
by Ruth Brown
HARRY, THE DIRTY DOG
by Gene Zion and Margaret Bloy Graham
DR XARGLE'S BOOK OF EARTHLETS
by Jeanne Willis and Tony Ross
WHERE'S THE BABY?
by Pat Hutchins